E. J. Donnel

Outlines of a New Science

E. J. Donnel

Outlines of a New Science

ISBN/EAN: 9783337208813

Printed in Europe, USA, Canada, Australia, Japan

Cover: Foto ©berggeist007 / pixelio.de

More available books at **www.hansebooks.com**

QUESTIONS OF THE DAY. NO. LVI.

OUTLINES

OF

A NEW SCIENCE

BY

E. J. DONNELL

NEW YORK & LONDON

G. P. PUTNAM'S SONS

The Knickerbocker Press

1889

Press of
G. P. Putnam's Sons
New York

PREFACE.

THE purpose of the following address was to meet a want developed in the present tariff controversy. I had not intended to publish any thing on the scientific aspect of the question until after the first revision of the existing tariff had been effected, when the public mind will resume its wonted calm, enlightened by the long debate and by the practical experience of a step, however short, toward the emancipation of our national industry.

Every year's experience adds to the strength of my conviction that a larger and more enlightened statesmanship is required to save society from revolutionary influences by providing free play for the tremendous forces which science has added to human industry. More and more deeply am I impressed with the belief that delay is dangerous.

Limited as I was to a single address, my statement of the science was necessarily con-

densed into the fewest words possible ; yet I
flatter myself that I have succeeded in making
it sufficiently clear to be easily understood by
all intelligent readers.

It is self-evident that, as the object of all
human industry, in an economic sense, is the
production of value, the law of value must
be the primary law, the supreme law—in other
words, the synthesis of economic science.

I hold it to be equally self-evident, when
clearly and etiologically stated, that exchangeability—commercial exchanges—is the supreme law of value.   This is very far from being a mere abstract truth to be realized in the
distant future.   It is the most pressing living
issue of our time.   The old theory that labor
employed in production is the law of value,
has caused incalculable evil in giving a wrong
direction to legislation and in breeding false
theories in sociology.

On closer examination it will be found that
all social science is ONE, having the same synthetic basis.   Economics constitute one of
the aspects of this new science ; not, as heretofore supposed, a discrete department of it.

This opens wide the door to a comprehensive study of the constitution of the human mind. Here it will be found that the law of sex prevails and is as potent as elsewhere. It is known to science as positive and negative forces. It is from the interaction of these that all ideas, and, consequently, all individual and social progress, are begotten. The same law, which is known among merchants and generally in the ebb and flow of social dynamics as action and reaction, will be found to be the only safe guide for the merchant as to the past and future of the markets ; to the statesman as to the legislation required to moderate extreme expansions and contractions which are so fruitful in suffering ; to the historian and philosopher it will explain the past and help to a general knowledge of the future. It will be found that the rise and fall of prices which take place on our exchanges almost every day, and sometimes every hour, are produced by a cause that is potent in all departments of human affairs. The human mind is always moving from an extreme in one direction to an extreme in the opposite direction. At one extreme is optimism, at the other pessimism. The former is characterized by bold-

ness, self-reliance, and enterprise ; the latter,
by absence of these qualities.

It will not be difficult to trace these movements from the hourly pulsations of the Stock
Exchange to the actions and reactions in European politics, so sharply accentuated since
the French Revolution ; and to the great
movement of two thousand years from the
times of Æschylus, Plato, and Aristotle, to the
times of Shakespeare, Bacon, and Newton.

To the pessimist these movements seem to
be circular, but they are in reality spiral. We
have already entered upon a negative era
which promises to last at least a thousand
years, but we are equipped for it with an
amount of positive knowledge so great that it
will differ very widely from the negative era
ending with the seventh century.

It is not necessary to emphasize the importance of a thorough knowledge of the law of
wages. There is a widespread delusion that
wages can be regulated without regard to the
amount and value of the work done, either
to the wage payer or to society. Trades-
unions insist that inferior workmen shall be
paid as much as the best. The law of wages

will teach them that this is done at their own expense, not at the expense of the wage payer. Seeing that production is the law of wages, the way to higher wages is cleared from all obscurity. It is demand for commodities that increases the productiveness of labor ; and excessive demand can only come from extreme activity in commercial exchanges. When there is work and wages for two workers and only one worker to be had, the offer of double wages will sooner or later find one who will do it all, provided it is under a social and political system that leaves all avenues open to personal ambition. When increased production first takes place, the larger share of the increased profit is obtained by the wage payer ; afterwards it is for a time pretty evenly divided between the wage payer, wage earner, and consumer ; finally it is all divided between the two latter. This is the natural order of evolution. Discriminating taxation, such as our so-called protective tariff, acts powerfully in the interest of the capitalist, sometimes giving him a monopoly of the whole profit from the increased productiveness of labor.

It is seen that the whole secret of industrial prosperity exists in commercial exchanges, which create and sustain active, continuous demand. This truth is thoroughly practical. " It is not a question of maxims, but of markets," —markets everywhere all round the globe.

If the New Science should be taken up by others more competent than I am, and adequately treated, I will not return to it. If not, I will resume the subject on a more extended scale.

E. J. DONNELL.

A NEW SCIENCE.

LECTURE DELIVERED BEFORE THE REFORM CLUB
OCTOBER 5, 1888.

Introduction.

WHEN the two great national parties divided on the tariff question and both the politicians and the people became thoroughly interested in it, I thought my participation in it would no longer be necessary or useful. I was glad to think so; for I had learned that the work of exciting popular interest in a question about which the great majority were indifferent, was any thing but pleasant. It was like addressing an audience at once deaf, dumb, and blind. It was very trying to the temper.

The great ability with which the debate has been conducted by the reformers, both in and out of Congress, seemed, for a time, to leave nothing to be desired. The accumulation of facts and statistics was, to appearance, overwhelming. But it soon became apparent that something more was required—that without knowledge of the laws that govern facts, there could be no end to the debate. It was found that facts could be manufactured to order. This industry is now more flourishing than any other in this country. Without any protective tariff it pays the highest wages. Whether it will be profitable to the capitalists depends on the election in November.

It is now about twenty-five hundred years since the belief that nature is governed by uniform laws first found expression in extant European literature. This belief laid the foundation for all subsequent progress in scientific discovery. In social science the work hitherto done has been mainly in the way of digging for and shaping materials; very little that can be called construction has been accomplished, and still less in accordance with a complete architectural plan.

This is the work to which I am now about to address myself. I cannot hope to accomplish more than a beginning—a mere outline, which may be filled up hereafter. I ask your close and patient attention. I must proceed rapidly to do even this much in one evening.

What Constitutes a Science.

The progress of knowledge changes the meaning of words when it does not result in substituting new words for the old ones. It sometimes happens that, when the old words are retained notwithstanding the enlargement of the thought and without new definitions, progress is retarded and truth obscured. Thought and language act and react on each other—obscurity in either causes un-certainty in the other.

It is easy to perceive that, after Newton made astronomy a science by the application of the law of gravitation, an enlargement, if not an entire change in the meaning, of the word Science became necessary. We could no longer regard mathematics as the ancient Greeks, who considered it *The* Science *par excellence.*

The term science can no longer be applied to mathematics and to astronomy as embody-ing the same thought in both cases. Applied to mathematics it could only mean exact knowledge, whereas applied to astronomy it means law—that is, a force or potent cause governing homogeneous phenomena.

A science, thus understood, is somewhat analogous to the solar system—the potent force answering to our sun, and the homogeneous phenomena being the movements of all that part of the universe governed by it; the governing body being positive and the governed negative.

It is admitted by all competent authorities that, if we would be successful in the study of social science, we must follow the methods that have been successful in physical science.

Social Science Based upon the Constitution of the Human Mind.

I hold it to be self-evident that, if there be a synthetic basis for a social science, it will be found in the constitution of the human mind. I hold it to be demonstrable that there is such a basis, not occult, but obvious, indisputable, and self-evident.

No animal when born into the world is more helpless than man. It is only through the constant care and attention of others that his life can be preserved ; and it is only through contact with other men, and mental commerce with them, that his capabilities can be fed and stimulated into activity. If isolated, he could not possibly rise much above the four-footed beasts, and not to a level with domestic animals, but remain on the low level of the isolated animal—the wild beast.

Though man is probably endowed with a natural capacity for endless progress, it is morally certain that he owes to society—to intercourse with other men—every thing that distinguishes him from the brute creation. To this cause he is indebted for all his faculties.

Wherever two human beings meet, whether as antagonists or in friendly commerce, there is an instantaneous draft on the vital forces of both, which, through the interaction of positive and negative forces, stimulates and fertilizes mind. When two human beings first look into each other's eyes, progress begins. When contact takes place in antagonism—in war,—the mental development is most rapid, but it is never permanent unless it is supplemented and superseded by friendly exchange of services—commercial exchanges.

The histories of the Roman and Mahometan conquests are striking proofs of this truth. In both cases the development of power was wonderful, but not more so than the utter collapse, stagnation, and ruin that followed.

Commercial Exchanges the Primary Law of Economic Science.

Friendly commerce is always a constructor—a builder ; war is always a destroyer, and its advantages are necessarily temporary and fleeting. Yet so all-powerful is the force that draws men toward each other, that if they are prevented from friendly commerce they will inevitably come together in destructive antagonism.

Every barrier raised against commerce is an incentive to war ; every removal of such barriers is in the direction of peace, prosperity, wealth, and happiness.

Irrepressible Conflict between Commerce and Militaryism.

The history of European progress is a record of the struggle for ascendancy between militaryism on the one side and free commerce on the other. It was the opening of the Egyptian ports by Psammetichus that started the commercial and intellectual development of Greece ; it was the commercial supremacy of Athens that gave that city its intellectual supremacy and made her the fecund mother of European culture.

Under the Roman Empire commerce was subordinated to militaryism, and the result was utter ruin to the whole social fabric. The first dawn of a new era was the development of commerce and manufactures in the south of France. As early as the eighth century this movement came into contact and afterwards into commercial relations with the Mahometans of Spain. This developed an extraordinary intellectual movement in both. So powerful and aggressive did this become in the great commercial cities, Toulouse, Carcasonne, Beziers, and Narbonne, that the conservative powers which dominated Europe thought it necessary to their own safety to extinguish it in blood and ashes. This transferred the movement to Italy, where under the influence of the commercial cities arose what is known as the Renaissance.

It was the Provençal poets who furnished for Dante the models for his immortal poem.

The struggle between these antagonistic forces has never ceased for a moment ; it is as active to-day as it has ever been. Notwithstanding the enormous increase of international commerce, militaryism keeps pace with it. The continent of Europe trembles under the tread of armed men drilling and manœuvring. All the profits of commerce and the resources of science are employed to increase the destructive powers of war. Of course there is a tariff war also in full activity. Hostile tariffs always either precede or follow the clash of arms.

The fate of Europe depends upon the issue of this struggle. If commerce should be again subordinated to militaryism, as in ancient Rome, the consequences would be similar.

The same struggle has begun here. Our present tariff originated in our civil war. Such laws could not by any possibility have been enacted during peace. In order to maintain this system it is found necessary to foment international jealousy and discord.

The Senate of the United States, whose

most important function is to aid the Executive in making treaties with foreign powers, has practically abdicated its functions. Worse still, it has denounced and condemned negotiation as a means of settling international disputes. Thus, it not only declines to discharge a duty which the Constitution imposes upon it, but it undertakes to denounce the Executive for attempting to discharge his part of the Constitutional mandate. This has a significance which I commend to the impartial consideration of the American people.

Political Economy the Science of Man.

It is that aspect of social science that is known under the term economics with which we are now to deal ; but, when we remember that every faculty man possesses, physical, intellectual, and moral, has an economic value, it will be seen that economics, in reality, includes every thing concerning man ; and may be named either sociology, or political economy, or, better still, The Science of Man.

Viewed from the scientific standpoint, political economy or economics is based upon what is known as value.

Economic Origin of Value.

Economic value has its origin in exchange-ability exclusively. Things that cannot, from any cause whatever, unless the cause be temporary, be exchanged, though they may be useful, can have no economic value. No civilized man could by any possibility provide for his own wants without assistance. No man is capable of providing himself with food, clothing, and shelter with such expertness as to leave a surplus.

A man may be an expert hunter, or farmer, or carpenter, or shoemaker, but he cannot be all of these, nor any two of them. All experience proves that the highest perfection can only be attained by absolute concentration of all the faculties in a single direction. This is known as the law of the division of labor, which, through commercial exchanges, results in value and all that is known as wealth, which is surplus value.

It is probable that there is not one sane healthy human being among the fifteen hundred millions inhabiting our globe who could not do something for us more to our profit than we could do it for ourselves ; none anywhere with whom free commercial exchanges would not be profitable to both parties. No individual, no community, no nation can do every thing necessary to civilization in the best way. The individual who is excluded from exchange of services with his fellow-man sinks to the condition of barbarism, and the nation so excluded, however extensive its territory and varied its climate and productions, will, sooner or later, cease to progress, or it will disintegrate from internal convulsions.

Exchangeability the Origin of Wealth and Creator of Industry.

The principle of commercial exchanges, as I have stated it, has the same relation to the science of political economy that the law of gravitation has to astronomy. Not only is exchangeability the source of ·all wealth, it is also the creator of productive industry.

When demand precedes production, we have prosperity, with confidence in the future, and buoyant hope ; when production precedes demand, there is doubt and anxiety, with frequent disappointment, and finally stagnation and suffering.

When questions of taxation and revenue, of banking and currency, of international relations and diplomacy, of justice and equality between nations and individuals, or any other question whatever concerning individual and social well-being is studied and resolved in the interest of the freest, easiest, cheapest, and most extensive commercial exchanges with all mankind, it is most fruitful in happiness, because it is assuredly guided by the light of science.

You can perceive that the conclusions of science place the department of commercial exchanges in the vanguard of the industrial army. If you will look around you and study history you will find that nature gives abundant testimony to this. It is around the department of commercial exchanges that great wealth has always accumulated. It is in this department that governments indulging in extravagant expenditures find their greatest resources. The wonder is that commerce can live under such burdens. There can be no stronger testimony to its natural profits. Some governments kill the goose, but others, more cunning, content themselves with plucking it. When Bismarck introduced the protective system into Germany in 1879, his main argument was taken from this country, which, he said, raised an immense revenue *without the people knowing it.*

Government interference, as in this country at present, may cause the accumulation of great fortunes by a few producers, through monopoly, but even in this country the greatest fortunes have been made by expediting and cheapening transportation.

Historical Illustrations.

Some of the wealthiest communities in the past have been engaged almost exclusively in making commercial exchanges. Witness Holland. Witness the Hanseatic League, which so stimulated the productive industry of Germany for two hundred years, increasing the well-being of the masses of producers without doubt, but, at the same time, accumulating the great masses of wealth in the hands of the merchant and bankers.

The same was true of the Italian cities, Milan, with her two hundred thousand persons employed in manufacturing industries, was, no doubt, prosperous and happy ; but it was in Venice and Genoa that great wealth and power concentrated, most probably with less happiness.

Modern political economy had its origin in the contrast between the economic conditions of the peoples of Holland and France in the first decade of the seventeenth century.

About the year 1610 *Antoine de Montcretien,* a Frenchman, visited Holland. He was a man of rare gifts, a poet, a philosopher, a man of keen sensibilities and human sympathy. He was also an ardent patriot, and, withal, thoroughly practical. When he returned to France he established an iron manufacture of various utensils, and wrote a treatise on political economy, the first book ever published with that title. In that treatise I find the following passage with regard to Holland :

"Never has a state done so much in so little time. Never have people so weak and obscure had such lofty, such clear, and such sudden progress. The heavens cover no people so barbarous that it does not communicate with them. There is no corner of the world so distant it does not reconnoitre it ; nothing so secret it does not bring it to light. All lands are open to it through the sea. This marvel accuses our indolence, I will not say cowardice, the French nation is too brave.

" This wealth, so great and so promptly amassed that it seems even to those who possess it that it has come to them in a dream, taxes us with carelessness. I would be wrong to say want of industry, for no nation equals us on that point, either by sea or land.

" What will I conclude then after having collected my wits, dazed with admiration ? That having come at the end of the centuries it has profited by the whole experience of the past, wishing to confound in all others the hope of the future, so that to French industry it has united English management. If I wished to leave to posterity a picture of the utility of commerce, as formerly Homer did of peace and war on his famous buckler of Achilles, I would describe here the cities of Amsterdam and Magdeburg in the condition that they were twenty-five or thirty years ago, and on the other that in which they are now, crowded with people, overflowing with merchandise, full of gold and silver."

History Repeats Itself in This Country.

We are not dependent upon historical data for proof of the supreme importance of the department of Commercial Exchanges to the whole industrial organization.

Contemporary facts and experience are the only true interpreters of history. When we understand the present we will comprehend the past and, I will add, the future also. We may be quite sure that the same laws that govern the present have governed the past and will govern the future. Nor is it necessary to go to a distance in search of truth. If I am asserting a scientific truth, its verification can be found everywhere in all human societies. Every capable observer and thinker can find its verification, not only within his own personal observation, but within his own experience. Science must rest on universals or it is not science at all.

Our own country, though debarred from foreign commercial enterprises for nearly thirty years, is not wanting in signs of prosperity, so far as accumulation of wealth is concerned. Whence comes this? I will tell you.

Within twenty-five years we have constructed over one hundred thousand miles of rail-roads. In doing this we have drawn from Europe a large part of her surplus capital. Of course this involves a heavy accumulation of debt. This undoubtedly tempts us towards the well-known evils which always attend speculation with abundant credit. It also causes us to overestimate our riches. Our debts we count as wealth, not only in our minds, but they are so counted in our census.

This large indebtedness will be felt severely in our foreign exchanges at no distant day. It would be so felt now, but that we are still borrowing all we need, if not all we ask for. The danger of the situation is that it places us in some degree at the mercy of our creditors. So long as Europe is able and willing to lend to us, all will go well. If Europe should either need her capital at home or lose confidence in our securities, the consequences might be very serious. The excess of our exports over our imports of late years, notwithstanding our large borrowing, shows both our heavy indebt-edness and our lack of profitable foreign trade.

Perhaps, after all, the most alarming feature in the situation is that we have a great party, composed of half our people, whose leaders are constantly boasting of the balance of trade being in our favor! Just think of it, half of our people still holding to the mercantile system, or balance-of-trade delusion, which desolated Europe in the sixteenth and seventeenth centuries.

But the benefit we derive from this is immense. I have given much thought to this subject, and the conclusion I have arrived at is that the much greater part of our increased wealth and prosperity, real and apparent, such as they are, during the last twenty-five years, has its origin, primarily, in the extension of our railroad system. This has brought into cultivation a vast territory previously unproductive. It has drawn to these lands immigrants from all countries. The railroads have extended their feeders to every city and near to every village. These railroads must have business at any price. It has not been a question as to what the railroads could afford, but whether they could carry freight at a rate that would enable the farmer to produce and compete with other farmers of whatever country. The railroads could not stand still,—it was do or die! They accepted the situation. The result has been wonderful. Railroad charges have been reduced far below those of any other country in the world, yet with profit to the roads. To many people this was a genuine revelation. Mr. Vanderbilt, when asked at how low a rate railroads could afford to carry freight, replied: " That is what no man has yet discovered."

The contribution of our railroads to the extension and profits of our commercial exchanges, domestic and foreign, is simply incalculable.

I do not think our protected industries, so called, have added any thing whatever to our national wealth. The fortunes accumulated by the protected capitalists have been taken from the self-supporting industries. Imagine what our condition would be to-day if our foreign exchanges had been as free as our interior exchanges. If it had not been for the rapid extension of our railroads, utilizing our vast natural resources, our tariff restrictions would have ruined our country more than twenty years ago. It may be that our restrictive tariff has forced American enterprise into railroad building to a greater extent than would otherwise have been the case. It seems to have been the only way in which American industry could save itself from ruin. In the nature of things, such expedients can only be temporary. There are evils attending them that accumulate silently like the germs of disease, which are sure to prove fatal in the end, if not destroyed. The evils are widespread and various. I cannot enter upon details, nor would it be suitable to my present purpose. The question at issue presents itself in its largest proportions and with most imperative demand for a solution.

Importance of the Present Crisis.

We are traversing a crisis not second in importance to any in our history. Its importance is greatly enhanced by the fact that it is not confined to this country alone. It is felt in some degree in all the advanced industrial nations. Everywhere capital and labor confront each other with suspicion and anger. In Europe discontent is in part political ; here it seems to originate in economic causes exclusively : in both it tends to become social and Revolutionary. This crisis differs from all others of which we have any record in this : it does not originate in want or suffering ; the markets everywhere are glutted with commodities.

During the last hundred years physical science and mechanical invention have advanced with gigantic strides, so that the productive powers of labor have been increased far beyond the wildest dreams of our forefathers. This movement is led by the United States, where the productive powers of labor, in proportion to numbers, in our staple industries, are from two to tenfold greater than in other countries. Why is it that this wonderful progress has not only not produced universal well-being, but has actually been a potent cause of discontent? I will tell you.

While physical science has advanced so rapidly, social science—in other words, Statesmanship—has either stood still or retrograded. Irregular progress of this sort wrenches society from its natural equilibrium. The energies of the people have all been directed to the rapid production of commodities, while the government has been almost equally active in restricting distribution and consumption. Excepting in the departments of finance and banking, our legislation has not only not made any progress in a century, but has actually retrograded a century. Politicians still quote the fathers on questions of taxation, forgetting that the great merit of our revolutionary fathers was in the fact that they tried to adapt their policy to the conditions then existing. Nor is this the worst. These gentlemen ask us to follow the example of England during the sixteenth and seventeenth centuries—the times of her ignorance,—and denounce as traitors people who are willing to accept the wisdom for which England paid so dearly.

In France, England, and Germany, under the protective system as under free trade, government action has been directed to the promotion of foreign commerce, by cheapening raw materials, by reducing the cost of transportation, by opening communications with all countries, by colonial expansion, and by treaties of commerce where such expedients are necessary. In the protectionist countries, as in the free-trade countries, the end aimed at is identical, though in the former it is seriously embarrassed by the self-contradiction of attempting to reconcile the false with the true.

In this country our legislation has been guided by the theory that foreign commerce is an evil ; that it is best for us to produce at any expense all that the country is capable of producing ; that wealth has its origin in production and not in exchangeability ; that it is beneficial in the end to give home producers a monopoly of the home market, even though it should result in a home monopoly ; that monopolies in raw materials should be maintained, though they render exports of manufactures impossible, and that this should be compensated by high protection to manufacturers.

Economic Delusions.

These ideas do not, or at least they ought not to, belong to the nineteenth century. They look like the survivals of pagan superstitions, which may be found to-day in all Christian countries. Three hundred years ago the opinion obtained very generally that all foreign commerce was a kind of piracy, and that trade generally was little better than pillage, in which the gain of one was the loss of another. Even as late as the last century, the great school of economists known as the Physiocratic School held that all wealth comes from land, and that neither trade nor manufactures produced any thing, but merely changed the place or form of commodities taken from land.

The Socialists assert that labor used in effecting exchanges is unproductive, and adds nothing to wealth, while it absorbs most of the profits. Hence their bitter denunciations of what they call the *Bourgeosie*, a word derived from free *Bourgs* of the middle ages.

This sounds strangely, but not more so than the statements of Ricardo that "no extension of foreign trade will increase the amount of value in a country," and that "as the value of all foreign goods is measured by the quantity of the produce of our land and labor which is given for them, we should have no greater value, if by the discovery of new markets, we obtained double the quantity of foreign goods in exchange for a given quantity of ours."

Ricardo's theories of value, of rent, and of wages are no better. They are all ideal conceptions which have at best but a remote connection with facts and experience. Yet these theories passed current for half a century, and made their author famous. They have at last reached the bad eminence of forming what is called the scientific groundwork of German socialism. I sometimes think that, in the passage of the human mind toward truth, it may be in the order of nature that it should have the experience of all possible forms of error.

J. S. Mill repeats and amplifies Ricardo's errors, adding many of his own, but also many valuable suggestions. The main fault of his work is that he makes no attempt to distinguish between primary and secondary causes—between the potent cause and mere influences. In other words, his method is empirical, and lands the reader in a labyrinth without a thread of law to guide him. His method, if it may be called a method, is the substitution of exhaustive analyses for scientific generalization. The same method, with the addition of a microscope, constitutes the pseudo science of *Karl Marx.*

The world owed much to Adam Smith, but it has paid for the benefit much more than it was worth.

Manfred and *Faust* sold their souls for what they most desired. The completeness with which Smith annihilated the baneful theories of the mercantile system which had cursed Europe for centuries, excited such a lively sense of gratitude that a whole generation prostrated themselves on their knees before him, and, if they did not give up their souls to him, they did what is probably as bad—they abandoned their right to think.

It is this that makes great reputations a curse quite as much as a blessing. Adam Smith immortalized empiricism, *August Comte* eulogizes him for treating economics in this way, and for not attempting to treat it scientifically. *Comte* refers to Smith's essay on astronomy as evidence that he knew too well what constituted science to attempt its application to economics.

Danger of Accepting Authority for Truth in Science.

It is the perpetuation of this error through the potent influence of Smith's great reputa-

tion that has made of political economy a
case of arrested development. De Quincey was
not far wrong when he declared that no discovery had been made in political economy
in a century. Carlyle was entirely right when,
after reading, or attempting to read, the manu-
script of Mill's " Political Economy," he pro-
nounced it the "dismal science."

No real science was ever dismal, nor even
uninteresting, to the human mind. Science
means law. Every human soul hungers and
thirsts after a knowledge of law. Science
means light—lucidity. Where there is dark-
ness you may be sure there is no science. A
writer states that *Karl Marx* is obscure because he is deep! Nonsense. Science brings
truth to the surface. When a truth is com-
prehended, obscurity ceases, not only in the
thought but in its forms of expression.

I will again quote from *Antoine de Montcretien*, the long forgotten but real founder of
modern political economy :

"Science is not a cutter of images, who
makes mournful and motionless statues to
place them on some pedestal. It is rather a
beautiful mistress who would render the hearts

of the men who love her ardent and restless for beautiful things."

You see the founder of modern political economy was an enthusiast ; as were all the early French economists, because they believed firmly that they were dealing with a science and were wooing nature to reveal one of her grandest secrets.

I would be derelict in my duty if I did not recognize the great merits of the economic works of Henry Dunning Macleod. He is the first writer, so far as I know, who has attempted successfully to treat political economy as a positive science. For reasons best known to themselves, contemporary writers rarely mention his name. If he has not finished the work, he has certainly begun it well. Nor would it be just to omit mention of the fact that nearly all the great economists have clearly recognized and stated the leading truth that, when put in its right relations to the *ensemble* of economics, will constitute them a science.

It was undoubtedly a scientific instinct that prompted Adam Smith to devote the three first chapters of his great work to a discussion of the division of labor on which the principle of free exchange is based. He thus recognized the supreme importance of this principle. *Bastiat* makes it the pivot around which revolves his matchless logic and lucid illustrations. Even Mill confesses that, without freedom of exchange there can be no science in political economy.

Other writers have gone still further, proposing to drop the term political economy and call it the science of exchanges. Surely there was but one step more needed.

This last step is now being taken, which will, in my humble opinion, put an end to the controversy, as verified Science always does.

Applications of Science to Present Needs.

Do not imagine that I am wasting your time in discussing mere abstractions. On the contrary, you will find that this is the most pressing practical question of the day. You know the troubled condition of our industrial system. You know that such conditions open the public mind to all sorts of false theories which bode no good to the future of our country and of the world. It has been found that every successful step taken towards constitut-

ing economics a science brushes away some
of the sophisms that constitute the various
kinds and degrees of socialism prevalent.

Witness the venom with which *Karl Marx*
refers to *Bastiat* and Macleod. He calls the
former, with affected contempt, the commercial
traveller of free trade, and the latter, the paid
agent of the capitalists! His disciples imitate
the dogmatic bitterness of their master. This
is the very spirit of discord and revolution.

Surely there can be no doubt as to what the
statesmanship of the future should be. The
let-alone policy has finished its mission.
The pernicious activity of our government
must be met by a policy equally positive in
the right direction, guided by science. Let
us begin by abolishing protection to monopolies in raw materials, and follow this by re-
moving such taxes as contribute most towards
increasing the cost of production, and bear
most heavily on the poor, who are least able to
bear the burden. This being done, we must
establish postal relations with all countries.
Our government should use all the resources
of diplomacy to remove barriers to freedom of
exchange everywhere.

Aristotle says it is the first duty of a legislator to know all about men. If our statesmen would only obtain this knowledge they would be masters of the whole science of statesmanship. Economic science is merely the practical application of this knowledge.

Objections to Freedom of Exchange.

I will now consider the objections made to freedom of exchange by the advocates of the restrictive system.

Ostensibly there is now but one objection pressed, almost to the exclusion of all others, viz., wages. The glaring encroachments of monopolies have almost thrown their advocates out of the field. The plan of the protectionist campaign is to rely upon the subtle, silent power of money and on loud, reckless appeals to the supposed ignorance of the multitude.

The Scientific Law of Wages.

Let us see what science says about wages.

The law of wages in all countries and in all localities is the degree in which labor is productive. This is as much the law of

wages as gravitation is the law of astro-
nomical science. Variations from it are only
apparent, like the oscillations of a planet.
Under all apparent variations the force of the
primary law persists and must prevail.

In discussing this question the difference
between time wages and labor wages—time
cost and labor cost—should never be lost sight
of for a moment. It is not merely in one industry and in one country, but in all industries
and in all countries, and localities, that this law
governs. You will find that where there is an
apparent exception it can be explained in accordance with the primary law. If there were
one single real exception it would not be the
true law of wages, because science, being
based upon universals, does not admit of
exceptions. There are secondary and consequently temporary causes that influence nomi-
nal wages at least. From time to time con-
flicts between labor and capital may cause
changes in the division of profits between
them.

So absolutely is production the law of
wages that increase of production per capita
is the only possible way of advancing the rate

of wages. A protective tariff does make employment uncertain and temporary, and thus sooner or later reduces the productive energies of labor, but under no circumstances can it advance wages. Though wages are advanced by increasing the profits of production, when these increased profits are derived from discriminating taxation, or any other kind of monopoly, they are not an addition to the aggregate wealth of the nation, but, on the contrary, reduce the ability of other industries to pay wages to at least the full extent of the artificial profits.

Neither legal enactments nor superabundance of cheap land can increase wages in any other way than by increasing the productive powers of labor. Within forty years our tariff has been doubled in average rates and much more than doubled in its protective features; within the same time the protective system has been entirely abolished in England; yet the percentage of advance in time wages in England has been quite as great as in the United States. In both cases the advance has been in proportion to the increased productive power of labor.

So with regard to land—Russia is not only highly protected, but has 50% more land in proportion to population than the United States ; yet it is the lowest wages country in Europe.

Like every other form of taxation, a protective tariff is a burden ; but much more than a revenue tax is it a burden on the wage earner and the farmer.

There is one way in which a part of the burden of such a tariff may be temporarily transferred from the back of the wage earner to that of the farmer.

The prices of agricultural products are not raised by our tariff, but are, on the contrary, lowered by it. There is a natural parity of value between commodities of which the price in money is merely the denomination. The prices of nearly all our agricultural products are fixed on a natural parity with the prices of manufactured goods in England. In this case, so far as food is concerned, the burden is transferred from the wage earner to the farmer, but afterwards it returns indirectly to the wage earner, because it reduces the ability of the farmer to pay wages.

Much has been accomplished by specialists to make the public mind accessible to more intelligent discussion of the wages question ; but much more could have been accomplished if the scientific law had been mastered and used as a guide—as the central light around which homogeneous phenomena could revolve ; in the light of which their movements could have been as easily understood as the movements of the heavenly bodies.

If you will place time wages and labor wages —in common parlance,—day's wages and piece wages—in the deadly parallel column, you will have an object lesson that the least educated laborer can understand. Place in one column the time wages, and in the other the value of what the labor produces, and you will understand why it is that time wages varies so greatly, not only in different countries but in different localities in the same country, being nowhere exactly the same in any two cities or states.

Look into the Census and also examine for yourselves the facts that come within your own observation. This is a scientific law and necessarily universal, and therefore capa-

ble of verification by anybody who has
the intelligence and patience to observe ac-
curately what is going on around him. You
will find that, in some localities as near to each
other as New York and Jersey City—fifteen
minutes apart,—in some departments of industry, time wages differ as much as 50%.

An equal difference exists between the Penn-
sylvania farm and the California or Oregon
farm, also between the Georgia or Carolina and
the Arkansas cotton plantations.

In every case, without exception so far as my
investigations have gone, you will find that
where time wages are highest, not only is the
profit obtained by the capitalist greatest,
but the percentage of the total product
obtained by capital is also largest, at least
in this country. I have no means of testing
the question as to any other country. I learn,
however, from good authority that the percentage of the total product accruing to capital
in England has been diminishing steadily for
years.

This much I can assert with perfect confidence : OF THE UNITED PRODUCT OF CAPITAL
AND LABOR, THE PERCENTAGE PAID TO WAGE

EARNERS IN THE UNITED STATES IS SMALLER THAN IN ANY COUNTRY IN EUROPE.

These facts make it absolutely certain that production is the standard of wages. It does more. [Mark it well, for it is of the highest importance.] It proves that WHEREVER AND IN WHATEVER INDUSTRY THE NATURAL STANDARD OF TIME WAGES IS HIGHEST THERE LABOR IS CHEAPEST. This is no mere assertion drawn out by the exigencies of controversy. It is a conviction forced by facts upon a mind little prepared for it.

This will explain how it is that English labor in most departments is cheaper than German labor, though time wages in the former are nearly double those in the latter.

I have found in both England and France, departments of industry in which time wages were higher than in this country, and, considering the purchasing power of wages, I should call them 50% higher.

It is in these very high time-wage departments that these countries defy all competition. This law holds good everywhere.

Every advanced industrial nation has specialties in which it has no need to fear competition. The country having the most numerous or most important specialties must inevitably be the world's industrial leader and master.

This is demonstrably the condition of our country at present. In the production of food, timber, metals, cotton and all the staple products of these materials and staple textiles made from free raw materials, we have the highest time wages and the cheapest labor in the world.

Natural Tendency to Division of Labor.

There is a natural tendency in industrial society to division of labor. It is only arbitrary laws that can prevent people from devoting themselves to that branch of industry in which they can be most successful. It is now proposed to abolish our tariff on wool. This will certainly cause a considerable export of our woollen manufactures. Will this prevent imports? By no means. We will have our specialties, and other nations will have theirs, and we will exchange services with mutual profit. France imports both silk and woollen goods from England, and England reciprocates. The French make both silks and woollens and many other things that no conceivable tariffs could prevent other nations from buying.

The standard of wages being the value of what labor produces, and the law of value being exchangeability—that is, demand,—which has its origin in an opinion or impulse or human necessity, we can see here an illustration of the synthetic unity of Economic Science, based upon the constitution of the human mind.

The department of wages is in immediate connection with the superior law of Economic Science—Exchangeability. Wages is the child of commerce. Without world-wide commerce and markets everywhere, steady, permanent employment is impossible.

There is an equilibrium which can only be maintained by universality. Droughts and floods may affect rivers and lakes, but cannot affect the ocean. Steady, permanent employment is the first condition of well-being among wage-earners.

Production being the standard of wages, this is also dependent on commerce, not merely as to the prices of the commodities produced, but also as to the quantity.

When the demand for labor exceeds the supply, labor becomes more productive, and will continue to become more and more so, so long as that condition lasts. When the supply exceeds the demand, effort and invention relax and labor becomes less productive.

You can see examples of this every day. On a large scale you can see it in nations. The nations having the most active and extensive commerce have the most productive labor and the highest time wages. England has the most productive labor in Europe, in leading articles, because, more than any other nation, she has pushed her commerce to the four corners of the earth.

In the seventeenth and eighteenth centuries, when she was under the protective regime, and unable to compete with a rival by peaceful competition, she made war on him without hesitation. As all the other nations, excepting Holland, were quite as much fettered by the protective system as herself, she had no trouble in outrivalling them in the open markets. Against free-trade Holland all her efforts were in vain until she determined to try war. This was the tragedy of Rome and Carthage over again.

It was cruel ; but all the same it made England the world's commercial leader.

This country alone can wrest the leadership from her, because we can take it and keep it either in peace or war.

Activity of commercial exchanges, with resulting demand, is the primary cause, but there are several secondary and consequently temporary causes of increase of productive powers.

1st. Abundance of cheap land within reach of sufficient markets.

2d. Equality of political rights furnishing the stimulus of unfettered ambition.

3d. Great wars are a powerful stimulant to increased production in nations that are able to preserve their commerce intact.

It was this that enabled England to survive the Napoleonic wars. It was during these wars that she first had her great success in the application of Science and Mechanics to industry. England alone of all the European nations pressed and extended her commerce during those wars. It is within the truth to say that she doubled the productive power of her labor in fifteen years.

One of the most important effects of our own civil war was the increase in the productive powers of labor in all the leading industrial nations. In England and in this country the increase was very great.

I have not named machinery, because man is the primary force and machinery is merely an instrument. When men have sufficient motives for doing so they never fail to invent or procure the necessary instruments to accomplish their purposes. Scientific accuracy would be impossible, if we permitted the least confusion of thought as between primary and secondary causes. It would be bad enough to confound permanent with temporary causes; but in the Science of Man, to put machinery on a level with its creator would be the last extreme of unreason.

Importance of the Scientific Method.

It is important that this wages question should be studied according to the Scientific Method—deductively as well as inductively, —because past experience proves that empirical treatment never reaches the end of any controversy involving principles.

This truth is strikingly illustrated in the history of the wages question.

When Cobden was agitating the tariff question in England, a Parliamentary Committee was appointed to investigate this very question with regard to the disparity between time wages in England and on the Continent. The report of the Committee was conclusive that, though time wages were much higher in England than in any part of the Continent, English labor was by far the cheapest. What is the reason that this did not end all controversy on the wages question ?

My answer is to repeat what I have already stated, in substance : no such controversy can ever be ended until we arrive at a knowledge of the Law—that is, Science.

Some thirty years after that Parliamentary report, we find Sir Thomas Brassey publishing a book ("Work and Wages"), in which he states that his father, who had constructed railroads in all quarters of the world,—in countries differing in time wages as much as five hundred per cent.,—found the cost of the work was about the same everywhere.

Indeed Mr. Brassey states facts that show the highest time wages to be the cheapest.

Now observe : Mr. Brassey writes as if he had made an original discovery. So far as the law of wages was concerned, he left the question just where he found it.

Professor Cairnes, a distinguished writer on political economy, criticises Mr. Brassey, accusing him, justly, I think, of having no confidence in his own assertion, that the cost of labor in different countries does not differ materially. Mr. Brassey's want of confidence is only such as must characterize all empiricists.

One would naturally expect that a distinguished economist, such as Professor Cairnes, would have at least suspected the existence of a stream of law flowing through such a surprising accumulation of facts. The idea does not seem to have occurred to him. Yet this law of wages is one of the simplest, most obvious, most easily verified, and one of the most important generalizations in economic science.

Now we have our government searching Europe for statistics about wages as if nothing was known about the subject. I do not blame the government. If blame is due anywhere it is to the economists. Perhaps even this would be unjust. Truth is often long germinating in the human mind, and requires the fulness of time before it bears fruit.

Piling up facts and statistics is the labor of Siysphus. On every new occasion the whole work has to be begun over again. The human mind is not very retentive of facts, though it may be attracted by them ; but laws, once understood, are never entirely forgotten, because they belong to the eternal nature of things.

The nearest approach to a scientific statement of the law of wages is by Professor Francis A. Walker. He states it thus, as a refutation of the old theory of a wage fund out of which all wages must be paid. " I hold that wages, in a philosophical view of the subject, are paid out of the product of present industry, and hence that production is the true measure of wages."

The defect of this statement is that it is unscientific. It makes the law of wages dependent on the postulate that wages are paid out of the product of present industry, which, as a matter of fact, is not true ; but even if it were true it has no place in the statement of a scientific axiom. Scientific laws are not

compound but simple ; they are not dependent
but absolute ; they are neither local nor con-
ditional but universal.

After making this statement, Professor
Walker proceeds to make an exhaustive statement of the various causes that influence
wages and production, admitting what he had
previously denied, that wages are paid out of
capital or by means of capital.

These self-contradictions are inseparable
from the empirical method. They leave the
reader's mind in a state of confusion and
doubt, and seem to have no permanent influence on economic science.

In his enumeration of the causes that influence wages and production, he omits all mention of the all-powerful cause—commmercial
exchanges. They produce all capital, create
and stimulate all surplus production, and pay
all wages. This is the real wage fund, not
fixed and limited, but an inexhaustible, living,
ever-flowing fountain. Where commercial
exchanges are in healthy activity, capital is
never wanting, but flows out, like arterial
blood from the heart, to all parts of the social
body.

When economists understand this truth as the one fundamental law governing all economic phenomena, they will have a light to guide them through the whole field of economic studies. Then their work will be luminous, consistent, and convincing. Multitudes will then read them eagerly ; whereas now to read them is a positive drudgery—always excepting the works of *Bastiat* and a few others.

Methods of Physical Science That are not Applicable to Social Science.

The slow progress of social science and especially of political economy, has arisen, primarily, from the dominating influence of physical science. The minds of economists have been preoccupied with this influence. From this bias has arisen their persistent search for physical causes for what are really mental phenomena having their origin in mental dynamics.

An instance of this bias which approaches the comical was the attempt of Professor Jevons to prove that sun spots are the cause of commercial revulsions.

It is demonstrable that the primary cause

of all expansions and contractions in trade is mental, and that all others are secondary. I cannot treat this question adequately in the time at my disposal. I will state briefly that the primary cause of these expansions and contractions is what is familiarly known as the law of action and reaction, which is universal and as active in mind as in matter.

Importance of the Law of Action and Reaction.

It is impossible to overestimate the importance of a comprehensive knowledge of this law, not only to the economist, the merchant, and the statesman, but to the philosopher.

1st. It illustrates the truth that value has its origin in the human mind as expressed by the word demand.

2d. Its universality shows the unity of all the sciences, the more so as it is inseparably connected with gravitation.

3d. Its potent influence on Social Evolution illuminates what is known as historical philosophy, which it advances to the threshold of science.

4th. Still more, it shows wherein universal laws are less potent in mind than in matter, the limitation of their action being possible through human intelligence ; thus demonstrating the freedom of the intelligent will.

I trust I have said enough to attract to this subject the attention of trained thinkers.

Another cause of the slow progress of Political Economy has been the exclusion from it of all questions of morals.

In a statement written by one of the students who attended the lectures of Adam Smith in the Glasgow University, occurs this passage : " In the last part of his lectures, he examined those political regulations which are founded, not upon the principle of *justice*, but that of *expediency*, and which are calculated to increase the riches, the power, and the prosperity of a state."

Nearly all the economists have followed this line ; they have treated political economy as a science that does not concern itself with or involve questions of morals. Mill repeats this idea with emphasis. It is quite in accord with the despotic dominance of physical science. It is only just to state that some French economists have always asserted that political economy is pre-eminently a moral science. H. D. Macleod adopts this view.

For the first time the moral aspects of our tariff controversy begin to receive attention from the masses of our people.

It begins to be seen that this is not a mere question of expediency. This is gradually imparting to the discussion a depth and earnestness heretofore unknown.

Ethics are indeed supreme. It is only questions of justice—the right and the wrong—the moral responsibility of man toward his fellowman—that take a deep hold upon the human heart. It is the one power that moves the world. So long as the public mind is unconscious of the moral aspects of political questions, affairs drift as if chaos were come again ; and thoughtful observers are driven to despair —to pessimism. It is only when the public conscience is aroused that hope revives and we become at once convinced that there is more than hope for humanity. Then and then only do we have real faith—faith that mounts to the skies and makes "life worth living."

This is not a mere question of the industrial supremacy of the United States, though that is involved in it. It is not a mere question of wages, though that also is involved in it ; it is not a mere question of the more or less increase of the national wealth, though that also is involved in it. It is something above and beyond all these—it is the question whether the principle of justice—free and equal justice—supposed to be the basic principle upon which our government was founded—is still strong enough in the hearts of the American people to enable them to live together in peace and harmony.

Twenty-two hundred years ago *Aristotle* wrote this golden sentence : " A firm state is that where people enjoy equality according to their merits ; and the secure possession of the property that belongs to them."

This sums up in one sentence the whole philosophy of government. The principle of justice and equality is to society the analogue of gravitation in physics. It holds society together. Its violation means disintegration and chaos. Assuredly this tariff question will never be settled finally until it is settled in a way that will satisfy an enlightened public conscience.

That our present system of taxation is outrageously unjust and unequal is certain. That the present position of the Republican party with regard to it is an open defiance of science, experience, and common sense, is obvious. How is it that a party that entered our national politics by helping to reduce by 20 per cent. the lowest tariff the Democratic party ever ventured to propose, and pass a tariff 57 per cent. lower than that now in force, has become the pensioned agent of the most aggressive and grasping monopolies known anywhere since the French Revolution ? It is important to get a true answer to this question, What has occurred once may occur again.

Two Parties Confront Each Other in All Ages— the Party of Industry and the Party of Pillage.

If you will read history carefully you will find that in the struggles, agitations, and revolutions there recorded, two parties—the same in all ages and countries—have confronted each other : one composed of people content to live by the fruits of their own industry ; the other aiming to appropriate to its own use the industry of others. To the honor of human nature, the first is by far the most numerous, but it is unorganized and absorbed in the labors by which it hopes to live and prosper. The latter, though a very small minority, by means of its wealth, organization, and energy, stimulated by selfishness, exercises an influence out of all proportion to its numbers. It belongs in reality to no party in politics, but aims at using all parties as opportunities offer. Neither has it any country, though it uses the catch-words of patriotism more loudly and persistently than others. It is always on the alert watching for opportunities when the public mind is preoccupied. It is a perpetual danger from which eternal vigilance alone can protect society. Civil discord, international jealousies, and war are its favorite opportunities.

This party may with perfect justice be called the modern *Condottieri.* They like to divide their services between both armies or parties so long as they are allowed the privilege of pillaging the country. It does not serve their purpose to be wholly on one side, because in case of defeat they run the risk of being exterminated. They have a special dread of having a clear line of demarcation drawn between themselves and the masses of the people whose rights and interests are threatened by them.

The importance of the approaching election is derived from the fact that the two great national parties are divided on fundamental principles more definitely than ever before in our history.

How the *Condottieri* tempted the Republican party little by little and step by step until they obtained complete possession of it, we all know.

It is but simple justice to say that the rank and file of the party is not responsible for this. *They* are very far from wishing to do injustice to their countrymen—to enrich themselves at the expense of the toiling masses. Nor do I wish to impugn the motives of a majority of their leaders ; but I do assert strongly that these have failed in their duty.

The most imperative duty of the statesman is to repel all attempts of individuals and combinations of individuals to make use of the government for purely selfish purposes. Such attempts will always be made when opportunity offers, not less, perhaps more, in a republic than in a monarchy.

Conclusion.

I must hasten to a conclusion. I will resume the leading points of my argument in as few words as possible.

1st. I have shown that all wealth—all surplus value—is the fruit of commercial exchanges, and that without such exchanges no surplus could possibly exist.

2d. I have shown that when the department of commercial exchanges is most active and vigorous and its sphere of activity most extended, it communicates health and vigor to all departments of productive industry ; when it languishes or is obstructed, disease sets in. When producers are waiting for demand, wage earners must wait for employment ; on the contrary, when demand seeks for commodities it will also seek for workers. Therefore, I assert that the supreme duty of statesmanship is to use all the constitutional powers of the government in clearing away obstructions to freedom of exchange, whether they spring from monopolies at home or lack of the means of easy communication with all countries, or obstructions raised by other less enlightened governments. By these means all social problems will be solved, and all beneficent work will be made easy by the immense increase of surplus value everywhere.

3d. I have shown that the enormous increase in the productive powers of labor in

the advanced industrial nations has created for
the economist, the statesman, and the philanthropist a problem which, with threats that
sound like the rumblings of a social earth-
quake, demands an immediate solution.   Pro-
tectionism and all other forms of socialism demand that we monopolize our abundance and
so get smothered in our own honey.   Science
and religion are at last reconciled and united
in demanding that the whole human race be
treated as brothers whose interests are all
united inseparably, with one nature, one origin,
one destiny.

4th. I have shown that the productive powers of labor per capita in this country, in
nearly all staple commodities, is greater than
in any other country as a whole, and that our
specialties in this respect are more numerous
and important than those of any other country.
I have shown that the cost of labor is in in-
verse proportion to the rate of time wages, the
former being lowest where the latter is highest,
and that this is a scientific axiom, true in all
countries and localities, and thus within the
cognizance of every man capable of observing
and reasoning adequately.   If all this be true,

it must soon put an end to all controversy on the wages question, and remove from our politics the basest and probably the most disgraceful species of demagogism with which we have ever been afflicted.

5th. I have shown that these facts place this country in the vanguard of civilization so far as it is expressed by superiority of industrial capability. As our productive powers are greater and the restrictions on our commercial exchanges more oppressive than in any other of the advanced industrial nations, our danger is greater and the demand for a scientific remedy more imperative than elsewhere. Our speculative railroad building, largely with borrowed capital, cannot go much farther. It already shows unmistakable signs of exhaustion. When from any cause Europe ceases to make new investments here we will have a commercial and financial crisis as serious as any we have ever known.

If I judge rightly the popular sentiment, we will have free coinage of silver at no distant day. There has been a great increase of private indebtedness during the last twenty-five years, owing to three causes : first, our tariff system, transferring wealth from the many, especially the farmers, to the few; second, our heavy payments of the national debt; third, extensive speculation in real estate. Debtors are many and creditors few. Debtors believe they are wronged by the demonetization of silver. It is in their power to force free coinage of silver, and I believe they will do it. It is of very little consequence whether the unit of value is gold or silver. The important question is, how will it affect credit?

Here again we come into close relations with the primary law which governs all departments of economic phenomena. When we remember that 95 per cent. of all the trade of the world is conducted on credit, and that money itself is the highest form of credit, we can see that it is impossible to overestimate the importance of credit to commercial exchanges, and the supreme importance of managing the silver question in a way that will strengthen rather than impair it.

6th. I have shown that our tariff system is unequal, taxing the many for the benefit of the few, which is unjust and immoral. No trust or combination to raise the price of any commodity artificially and permanently can be successful without the protection of a tariff on the foreign article. Trusts in articles that are largely exported, such as cattle and petroleum, etc., cannot artificially much advance the prices of the articles exported, without world-wide combinations, but they can reduce the prices of what they purchase, of which labor is always a part. All combinations of this sort look to two sources of profit : to reduce the prices of what they buy, and advance the prices of what they sell. By substituting one buyer and seller for hundreds, they have it in their power to accomplish in these directions every thing the law authorizes. By crushing out small proprietors this system tends to enslave the masses of the people by reducing them to the condition of dependent employees.

So powerful has monopoly become already that the once great Republican party thinks it good policy to defend trusts in the concrete, while condemning them in the abstract ; thus giving a sly wink to the monopolists, while throwing dust in the eyes of the people.

Is any thing more necessary to define the duties of the American statesman in the future ? All I ask of him is to lay aside the littlenesses which are unworthy his vocation, and to study this subject scientifically, both deductively and inductively. He need not seek for data at a distance. This is the science of man to which I invite him. Wherever man is found all necessary facts can be obtained. If he will only take with him the torch of science the work will be as full of pleasure as of instruction. He will feel that exalting pleasure which comes from *rapport* with the Infinite through contact with the universal laws. This work has been begun, and is in full activity. The present stage of our industrial development demands that precedent shall give way to science, and the science is getting ready. This is according to the order of nature, which in the progress of evolution provides for every new want by a new provision.

In 1881 I predicted that our present abominable tariff system would receive its death-wound before the end of 1887. I think President Cleveland's message gave it that wound. I will now venture on another prediction which I may not live to see fulfilled, but many of you will.

Before the advent of the twentieth century this nation will be emancipated, and in the undisputed industrial leadership of the world, for which nature has equipped it.

I know that some of the most gifted statesmen and economists of Europe foresee this, and look forward to it with fear and trembling. The consequences to the nations of Europe will depend upon their own folly or wisdom. If they will disband their four million soldiers, and put an end to their present chronic state of war, and throw down all barriers to freedom of exchange, it will be of incalculable benefit to them and to the whole world. Sooner or later they will be forced to this. Let us hope it will be sooner rather than later. Under our leadership, industrial activity, light, and civilization will be poured into the dark places of the earth, which are now filled with the habitations of cruelty.

Universal freedom of commercial exchanges is an indispensable precedent for all genuine intelligent social reforms. Without this condition precedent no radical permanent amelioration is possible. Without this as a foundation, and still more, with this truth discarded, the labors of the philanthropist, the socialist, the land reformer, and what not, will be all in vain. In the end they will discover that they have been building castles in the air.

THE END.

www.ingramcontent.com/pod-product-compliance
Lightning Source LLC
Chambersburg PA
CBHW022003190326
41519CB00010B/1374